CONTENTS

INTRODUCTION

Wind is a clean and inexhaustible source of energy that can be harnessed to produce power. Historically, wind power in the form of windmills has been used for centuries for such tasks as grinding grain and pumping water. Today sophisticated wind machines known as wind turbines are used in many parts of the world to convert the kinetic energy of wind to electric power.

Modern commercial wind turbines produce electricity by using rotational energy to drive a generator. These turbines are made up of a blade or rotor and an enclosure called a nacelle that contains a drive train atop a tall tower. Large wind turbines (producing up to 1.8 megawatts of power) can have a blade length of over 130 feet (40 meters) and be placed on towers some 260 feet (about 80 m) tall. Smaller turbines can

EXPLORING ENERGY TECHNOLOGY

WIND POWER

EDITED BY ELIZABETH LACHNER

Britannica Educational Publishing
IN ASSOCIATION WITH
ROSEN
EDUCATIONAL SERVICES

Published in 2019 by Britannica Educational Publishing (a trademark of Encyclopædia Britannica, Inc.) in association with The Rosen Publishing Group, Inc.
29 East 21st Street, New York, NY 10010

Distributed exclusively by Rosen Publishing.
To see additional Britannica Educational Publishing titles, go to rosenpublishing.com.

First Edition

Britannica Educational Publishing
J.E. Luebering: Executive Director, Core Editorial
Andrea R. Field: Managing Editor, Compton's by Britannica

Rosen Publishing
Amelie von Zumbusch: Editor
Brian Garvey: Series Designer/Book Layout
Cindy Reiman: Photography Manager
Nicole DiMella: Photo Researcher

Library of Congress Cataloging-in-Publication Data

Names: Lachner, Elizabeth, editor.
Title: Wind power / edited by Elizabeth Lachner.
Description: New York : Britannica Educational Publishing, in Association with Rosen Educational Services, 2019 | Series: Exploring energy technology | Includes bibliographical references and index. | Audience: Grades 5–8.
Identifiers: LCCN 2018006606| ISBN 9781508106258 (library bound) | ISBN 9781508106241 (pbk.)
Subjects: LCSH: Wind power—Juvenile literature.
Classification: LCC TJ820 .W5763 2019 | DDC 621.31/2136—dc23
LC record available at https://lccn.loc.gov/2018006606

Manufactured in the United States of America

Photo credits: Cover Bonerok/E+/Getty Images; cover, back cover, pp. 1, 3, 4-5 (background), 8, 16, 25, 34, 41, 43, 46 © iStockphoto.com/tolokonov; pp. 4-5 Janis Schwartz/Shutterstock.com; p. 6 © Elenathewise/Fotolia; pp. 9, 10, 14, 27 © Encyclopædia Britannica, Inc.; pp. 11, 22, 32, 36 (background) sakkmesterke/Shutterstock.com; p. 12 Alexei Belozersky/Shutterstock.com; p. 17 bmphotographer/Shutterstock.com; p. 18 Jeff Greenberg/Universal Images Group/Getty Images; p. 20 Hadrian/Shutterstock.com; p. 21 tony mills/Shutterstock.com; p. 22 Denver Post/Getty Images; p. 26 © IKO/Fotolia; p. 28 © Courtesy of the U.S. Department of Energy; p. 31 © Pizano/Fotolia; p. 33 Tony Moran/Shutterstock.com; p. 35 Doin/Shutterstock.com; p. 37 pedrosala/Shutterstock.com; p. 39 © Deepwater Wind Holdings, LLC; p. 40 Nick Upton/Nature Picture Library/Getty Images.

These wind turbines are part of a wind farm in western Kansas. Wind power has a long history and is now of growing importance as an alternative energy source.

be used to provide power to individual homes. Wind farms are areas where a number of wind turbines are grouped together, providing a larger total energy source.

Wind resources are calculated based on the average wind speed and the distribution of wind speed values occurring within a particular area.

The term "windmill," which typically refers to the conversion of wind energy into power for milling or pumping, is sometimes used to describe a wind turbine.

Areas are grouped into wind power classes that range from 1 to 7. A wind power class of 3 or above (equivalent to a wind power density of 150 to 200 watts per square meter or a mean wind of 11.5 to 12.5 miles per hour [5.1–5.6 meters per second] at a height of 33 feet [10 m] above ground) is suitable for utility-scale wind power generation, although some suitable sites may also be found in areas of classes 1 and 2. In the United States there are substantial wind resources in the Great Plains region and in some offshore locations.

Wind power is an alternative energy source. This means that the power of the wind can be used in place of other energy sources such as coal, oil, and nuclear reactions.

Fossil fuels, such as oil, coal, and natural gas, are available on Earth only in limited supplies, and they will run out one day. Fossil fuels also add carbon dioxide to the atmosphere when they are burned. Scientists believe that this contributes to global warming. Burning fossil fuels also adds other pollutants, such as carbon monoxide and nitrogen dioxide, to the atmosphere. Unlike fossil fuels, wind power is a renewable resource that cannot be depleted. It also has the added benefit of being virtually pollution-free.

Challenges to the large-scale implementation of wind energy include siting requirements such as wind availability, aesthetic and environmental concerns, and land availability. Public advocacy groups have raised concerns about the potential disruptions that wind farms may have on wildlife and overall aesthetics. For example, wind generators have been blamed for injuring and killing birds. However, experts have shown that modern turbines have a small effect on bird populations.

CHAPTER ONE

WHAT IS WIND?

Wind is a familiar element of weather, one that pretty much everyone has some experience with. However, many people don't know how to really define wind or what causes it. Simply put, wind is the movement of air across Earth's surface.

LOCAL WINDS

All winds, from gentle breezes to raging hurricanes, are caused by differences in the temperature of the atmosphere, by rotation of the Earth, and by unequal heating of the continents and the oceans.

Simple examples of the effect of temperature on wind are land and sea breezes. The sun heats land and water differently. Over oceans and lakes most of the heat energy is consumed in evaporation or is absorbed by the water. The air is not heated very much. Land, however, absorbs only six tenths as much heat as water does and evaporation is less. Hence air over land tends to receive more heat than does air over water.

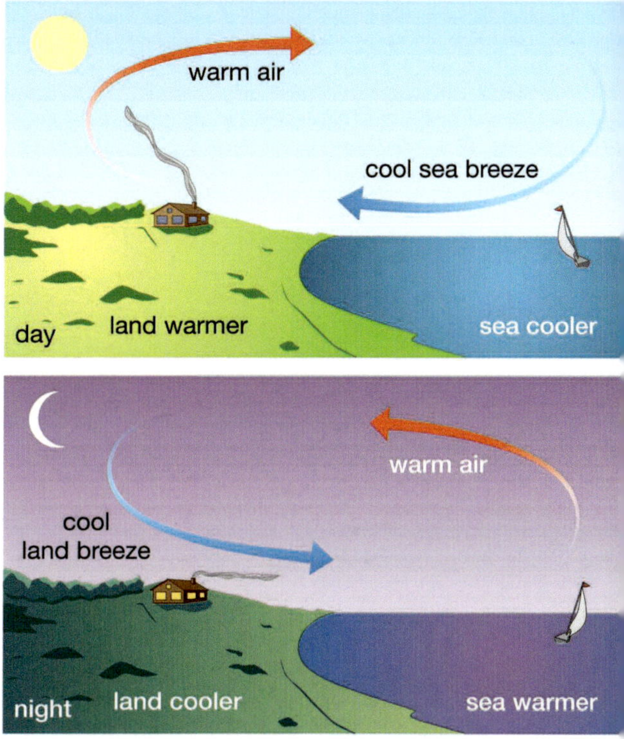

By day the land is warm and the sea cool. The warm air rises and the cool air descends. The breeze is off the sea. At night the land is cool and the sea warm. The breeze is off the land.

Heated air expands and becomes lighter. This starts to occur over land soon after sunrise. Air over the sea does not heat as rapidly, and so it remains denser, or heavier, than the land air. Its pressure near the Earth's surface is greater. At length this extra pressure starts it flowing landward, where it pushes the lighter land air up out of its way. This movement is called a sea breeze.

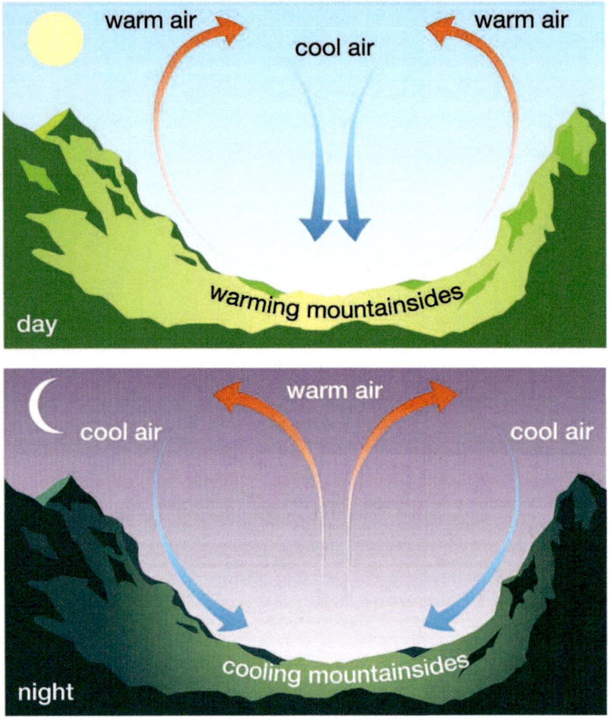

Day and night changes in heating mountains cause breezes. By day heated air rises along the warm slopes but descends into the valley. At night the circulation is reversed.

During the night the opposite action occurs. The land cools more rapidly than the sea and so does the air above it. At some time in the night the breeze is reversed and blows out to sea as a land breeze.

Similar day-and-night changes cause mountain and valley breezes. By day the greatest heating occurs along mountainsides and at the heads of valleys. Cooler, heavier air on the lower levels then blows up the valleys. At night this action is reversed.

PROOF OF WIND'S POWER

While they can be a power source, winds are also important for other reasons. These masses of moving air change the face of the Earth. Winds, like water, constantly wear down rocks and scoop out canyons. The powerful blasts of air carry away sand and soil. This is called wind erosion. In some places winds bury fertile soil under wind-borne sand. In others they lay down rich deposits.

ATMOSPHERIC PRESSURE AND HEAT

In general, all the winds near the Earth's surface are caused by differences in atmospheric pressure brought about by uneven heating. Air is pushed from regions where pressure is high toward regions where pressure is low. Weather maps make this plain, with the labels "high" and "low."

The notion that heating of air by the sun makes it lighter and decreases its pressure may seem contrary to experience. When we compress air in a tire

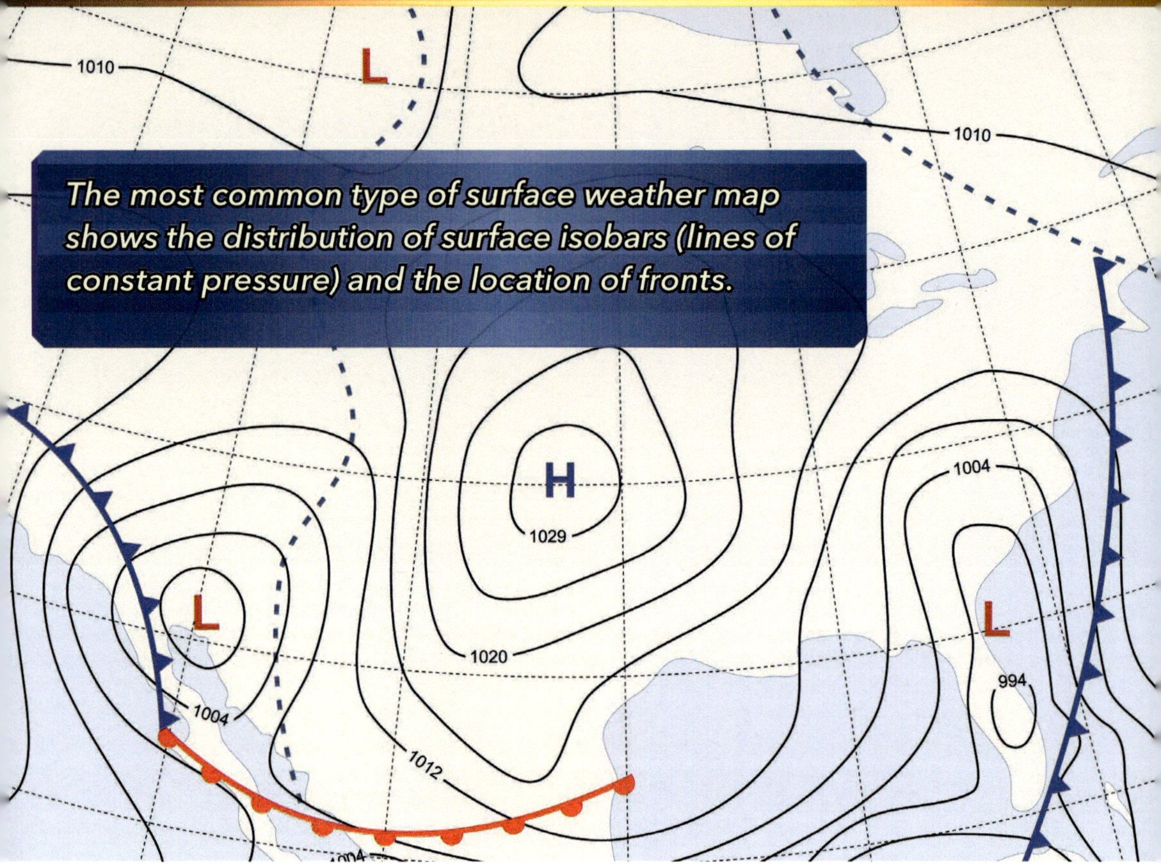

The most common type of surface weather map shows the distribution of surface isobars (lines of constant pressure) and the location of fronts.

pump, it becomes heavier and warmer. If we warm the air in a corked bottle, its pressure increases and may blow out the cork. In such cases the air is in a closed container. When air is free, heating expands it. There is less of it per cubic foot. It becomes lighter. The pressure of the atmosphere at any point depends in general upon the weight of the air above that point. Warm, light air has lower pressure than air that is cold and heavy.

When air near the Earth's surface warms and expands, it naturally tends to flow upward. We can say it "floats" up through the heavier, colder air around it. Or we can say that it is "forced" up by the greater pressure of the surrounding cold air. It means the same thing. There is an updraft of warm air that is not, strictly speaking, a wind. There is an incoming flow of cold air from the sides that is a wind.

In some cases a large mass of warmed air may expand outward from its center with force enough to push cold air back. But most warm-air movements along the surface are caused by the pressure of colder air.

PLANETARY WINDS

We have seen how local winds are caused by local changes in temperature. Similar air movements can occur on a worldwide scale in response to worldwide variations in temperature. These vast movements are called planetary winds, and they are driven by the circulation of air in atmospheric units called cells. The Northern and the Southern Hemispheres each have three mirrored cells based on latitude: one between the equator (0°) and 30°, another between 30° and 60°, and a third between 60° and 90°. The North Pole is 90° N, and the South Pole is 90° S.

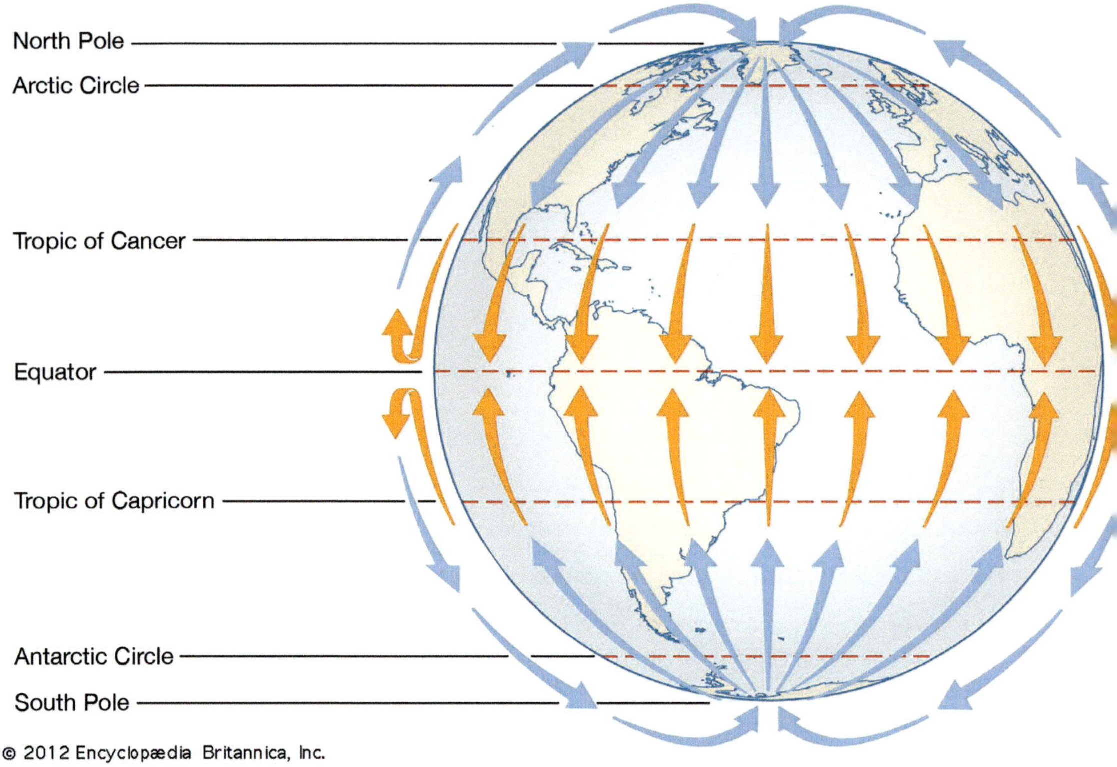

The sun's uneven heating causes winds around the world. The hot and light air in the tropics rises and flows toward the two poles. The cold and heavy air from the poles moves toward the equator.

The temperature variations that produce the planetary winds arise in various ways. The most important is the different effects of the sun's heat in the tropics and in the polar regions. Throughout the year, the most intensely heated part of the Earth is the tropical zone, which extends from 30° N to 30° S. The continuous heat from the sun in this area keeps the tropical air warm and light. Cooler, heavier air continually moves toward the tropical zone from the north and south, pushing the lighter and warmer air upward. Because of these upward currents, the middle of the heat belt—the equator—experiences little wind. As the ascending currents cool, their moisture condenses, drenching the zone continually with showers. Sailing vessels were often becalmed for weeks in these latitudes, prompting sailors to call the area the "doldrums," a term stemming from an old word meaning "dull" and "stupid."

Chapter Two

WIND POWER HISTORY

Windmills, together with waterwheels, were widely used from the Middle Ages to the nineteenth century during the course of which they were supplanted by steam engines and steam turbines. Windmills tap the energy of the wind by means of sails mounted on a rotating shaft. The sails are mounted at an angle or are given a slight twist so that the force of wind against them is divided into two components, one of which, in the plane of the sails, imparts rotation.

Though they continued to be used for pumping water in rural areas, wind turbines practically disappeared in the twentieth century as the internal-combustion engine and electricity provided more reliable and usually less expensive power. By

This farm near Avilla, Indiana, has an annular-sailed windmill. This style of windmill is common on American farms, particularly those in the Midwest.

the late twentieth century, however, they came back into use for electricity generation.

EARLY WINDMILLS

The earliest-known references to windmills are to a Persian millwright in 644 CE and to windmills in Seistan, Persia, in 915. These windmills are of the horizontal-mill type, with sails radiating from a vertical axis standing in a fixed building, which has openings for the inlet and outlet of the wind diametrically opposite to each other. Each mill drives a single pair of stones directly, without the use of gears, and the design is derived from the earliest water mills. Persian millwrights, taken prisoner by the forces of Genghis Khan, were sent to China to instruct in the building of windmills. The use of windmills for irrigation there has lasted ever since.

The vertical windmill, with sails on a horizontal axis, derives directly from the Roman water mill with its right-angle drive to the stones through a single pair of gears. The earliest form of vertical mill is

This post mill, shown here at Flowerdew Hundred, a plantation site in Virginia where it originally stood, is now located in Lubbock, Texas.

known as the post mill. It has a boxlike body containing the gearing, millstones, and machinery and carrying the sails. It is mounted on a well-supported wooden post socketed into a horizontal beam on the level of the second floor of the mill body. On this it can be turned so that the sails can be faced into the wind.

The next development was to place the stones and gearing in a fixed tower. This has a movable top, or cap, which carries the sails and can be turned around on a track, or curb, on top of the tower. The earliest-known illustration of a tower mill is dated about 1420. Both post and tower mills were to be found throughout Europe and were also built by settlers in America.

To work efficiently, the sails of a windmill must face squarely into the wind, and in the early mills the turning of the post-mill body, or the tower-mill cap, was done by hand by means of a long tailpole stretching down to the ground. In 1745 Edmund Lee in England invented the automatic fantail. This consists of a set of five to eight smaller vanes mounted on the tailpole or the ladder of a post mill at right angles to the sails and connected by gearing to wheels running on a track around the mill. When the wind veers it strikes the sides of the vanes, turning them and hence the

You can see the fantail on this windmill from Bremen, Germany. It is the smaller set of blades and sails on the back of the windmill.

track wheels also. These turn the mill body until the sails are again square into the wind. The fantail may also be fitted to the caps of tower mills, driving down to a geared rack on the curb.

The sails of a mill are mounted on an axle, or windshaft, inclined upward at an angle of from 5° to 15° to the horizontal. The first mill sails were wooden frames on which sailcloth was spread; each sail was set individually with the mill at rest. The early sails were flat planes inclined at a constant angle to the direction of rotation; later they were built with a twist like that of an airplane propeller.

LATER DEVELOPMENTS

In 1772 Andrew Meikle, a Scot, invented his spring sail, substituting hinged shutters, like those of a Venetian blind, for sailcloths and controlling them by a connecting bar and a spring on each sail. Each spring had to be adjusted individually with the mill at rest according to the power required; the sails were then, within limits, self-regulating.

In 1789 Stephen Hooper in England utilized roller blinds instead of shutters and devised a remote control to enable all the blinds to be adjusted simultaneously while the mill was at work. In 1807 Sir William

The Saxtead Green Post Mill, in Suffolk, England, features patent sails.

ANNULAR-SAILED WINDMILLS

The annular-sailed wind pump was brought out in the United States by Daniel Hallady in 1854. Its production in steel by Stuart Perry in 1883 led to worldwide adoption, for, although inefficient, it was cheap and reliable. The design consists of a number of small vanes set radially in a wheel. Governing is automatic and overspeeding is avoided by turning the wheel off the wind direction, thus reducing the effective sail area while keeping the speed constant. A typical pump can deliver about 10 gallons (38 liters) per minute to a height of 100 feet (30 m) at a wind velocity of 15 miles per hour (6.7 m/s). The best-known machines of this type are the so-called American farm windmills that came into wide use during the 1890s.

In this image from 1977, people install a wind turbine with an unusual annular-sail design.

Cubitt invented his "patent sail," combining Meikle's hinged shutters with Hooper's remote control by chain from the ground via a rod passing through a hole drilled through the windshaft. The operation was comparable to operating an umbrella. By varying the weights hung on the chain the sails were made self-regulating.

GENERATING ELECTRICITY

The development of the electric generator aroused some interest in wind as a "free" power source. The first windmill to drive a generator was built in 1890 by Poul la Cour in Denmark, using patent sails and twin fantails on a steel tower.

Adopting the ideas gained from airfoil and aircraft propeller designs, windmill designers and manufacturers began to replace broad windmill sails with a few slender propeller-like blades. In 1931 the first propeller wind turbine was erected in Crimea. From the 1940s, experimental twin-blade turbines were constructed in the United States and later in Scotland and France. In the Netherlands a few old-fashioned mills were adapted to generate electricity.

Interest in wind turbines for electricity generation was rekindled by the oil crisis of the mid-1970s. High initial costs, intermittent operation, and maintenance costs, however, have prevented wind turbines from becoming a significant factor in commercial power production. Today, wind turbines for electric power generation are most commonly propeller-type machines.

CHAPTER THREE

TODAY'S WIND TURBINES

Modern wind turbines extract energy from the wind, mostly for electricity generation, by rotation of a propeller-like set of blades that drive a generator through appropriate shafts and gears. The older term "windmill" is often still used to describe this type of device, although electric power generation rather than milling has become the primary application.

TYPES OF WIND TURBINES

There are two primary types of wind turbines used in implementation of wind energy systems: horizontal-axis wind turbines (HAWTs) and vertical-axis wind turbines (VAWTs). HAWTs are the most commonly used type, and each turbine possesses two or three blades

People have invented many ways to cause energy transformations. For example, wind turbines transform the mechanical energy of the wind into electrical energy.

attached to each turbine. VAWTs are able to harness wind blowing from any direction and are usually made with blades that rotate around a vertical pole.

HORIZONTAL-AXIS WIND TURBINES

Modern horizontal-axis wind turbines have from one to four metal blades. Each blade is twisted like an airplane propeller. An automatic governor rotates the blades about their support axis to maintain constant

Wind turbine

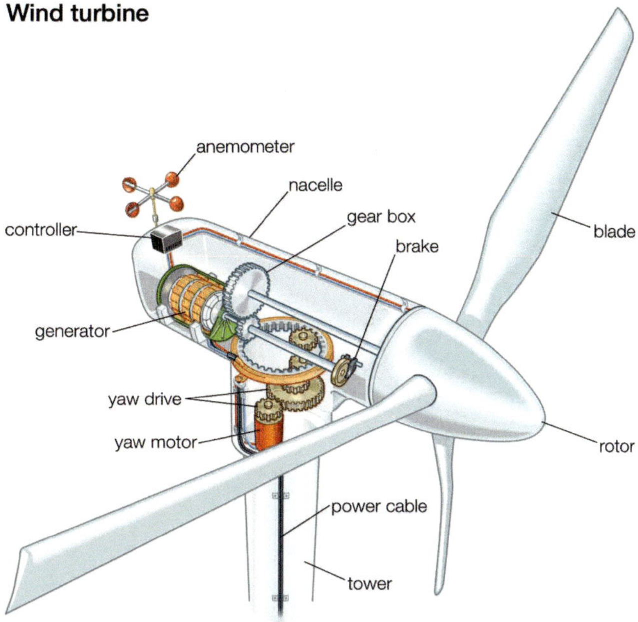

A wind turbine consists of many parts that work together to produce electricity. Three-bladed turbines like this one are the most commonly used design today.

generator speed. The Jacobs three-bladed windmill, used widely between 1930 and 1960, could deliver about one kilowatt of power at a wind speed of 14 miles per hour (6.25 m/s), a typical average wind velocity in the United States about 60 feet (18 m) above ground.

More recently, big horizontal-shaft, two-bladed turbines have been developed in the United States. The first such device, a unit equipped with a rotor measuring 38 feet (11.6 m) in diameter, was installed near Sandusky, Ohio, in 1976. Its power output was rated at 100 kilowatts. The most recent type of machine, first installed on the island of Oahu in Hawaii, has a rotor diameter of 400 feet (122 m) with its axis about 250 feet (76 m) above ground. Its output rating is 6,200 kilowatts at a wind speed of 29 miles per hour (13 m/s).

VERTICAL-AXIS MACHINES

Devices of this kind, which had not been used since the early Middle Ages, found a new application after the Finnish engineer S. J. Savonius invented a new type of rotor in 1922. Known as the Savonius rotor, it consists of semicircular blades that can be constructed from little more than the two sections of an oil drum, cut in half along its vertical axis and welded together with an offset from the axis to form an open S. An advanced version of this machine installed at Manhattan, Kansas, during the 1970s generated five kilowatts of electric power in a 27-mile-per-hour (12 m/s) wind.

The Darrieus VAWT, which uses curved blades in a curved arch design, became the most common VAWT in the early twenty-first century.

The most recent vertical wind turbine is based on a machine patented in 1931 by the French engineer G. J. M. Darrieus. Its two blades consist of twisted metal strips tied to the shaft at the top and bottom and bowed out in the middle similar to the blades on a food mixer. A Darrieus turbine with aluminum blades erected in 1980 by the Sandia National Laboratories in New Mexico produced 60 kilowatts in a wind blowing 28 miles per hour (12.5 m/s). Turbines of this variety are not self-starting and require an external motor for start-up. Several models of Darrieus turbines have been built since the construction of the Sandia unit.

LIMITATIONS ON WIND POWER

Not all the kinetic energy of the wind can be extracted, because there must be a finite velocity as the air leaves the blading. It can be shown that the maximum efficiency (energy extracted divided by energy available in the captured wind area) obtainable is about 59 percent, although actual wind turbines extract only a portion of this amount. Currently, the maximum efficiency obtainable with a propeller-type windmill is roughly 47 percent; this occurs when the propeller-tip speed is between five and six times the wind velocity. For a given rotor speed, it drops rapidly as the wind velocity decreases. The power obtainable varies as the square of the rotor diameter and the cube of the wind velocity. Thus the theoretical maximum energy obtainable from a rotor with a diameter of 98 feet (30 m) in a wind with a speed of 31 miles per hour (14 m/s) would be about 690 kilowatts. If the wind speed decreases to 16 miles per hour (7 m/s), the theoretical maximum drops to about 86 kilowatts. At this lower wind speed, it would require more than 17,000 wind turbines (with rotors of 98 feet [30 m] across) operating at an efficiency of 40 percent to match the output of a single large one-million-kilowatt central power station. When these limitations are coupled to the need for suitable sites with steady winds, it becomes apparent that wind turbines alone will not play a major role in meeting the power demands of an industrialized nation.

WIND FARMS

A wind farm is a cluster of wind turbines (up to several hundred) erected in areas where there is a nearly steady prevalent wind. Such areas generally occur near mountain passes. Wind farms comprising propeller-type units have been set up in Hawaii, California, and New Hampshire. Capacities range from 10 to 500 kilowatts per unit. During 1984 the total output of all US wind farms exceeded 150 million kilowatt-hours; the

Wind farms can be a valuable source of electric power in places where the climate is favorable.

entire output was fed into the electric utility network. Though seemingly substantial, this amounted to less than 1/100,000 of the total electric power generated in the United States.

By the early twenty-first century, wind was contributing slightly more than 1 percent of the world's total electricity. Electricity generation by wind has been increasing dramatically because of concerns over the cost of petroleum and the effects of fossil

WIND POWER IN THE UNITED STATES

The U.S. Energy Information Administration estimates that about 5.6 percent of the electricity used in the United States in 2016 was generated with wind power. In fact, it was second only to hydropower among the renewable energy sources for electricity production. Texas leads the United States in the production of wind energy. Most of the state's wind turbines are located in the Panhandle and in the Trans-Pecos region. Not all though: the Horse Hollow Wind Energy Center is spread across some 50,000 acres (20,000 hectares) near Abilene. Along with Texas, Iowa, Oklahoma, California, and Kansas all produce significant amounts of electricity using wind power.

fuel combustion on the climate and environment. From 2004 to 2009, for example, total wind power increased from 47,693 to 159,213 megawatts worldwide. By 2009 the United States possessed the most installed wind capacity (35,159 megawatts), followed by China (26,010) and Germany (25,777), while Denmark generated the largest percentage of its electricity from wind (20 percent). The wind

This offshore wind farm is located in the Baltic Sea, off the coast of Denmark. The country is a leader in the production of electricity using wind power.

power industry estimates that the world could feasibly generate 12 percent of its total electricity from wind power by 2020. Various estimates put the cost of wind energy between three and twelve cents per kilowatt-hour, depending on the location. This is comparable to the cost of fossil energy. (The cost of coal-generated electricity is estimated at four to eight cents per kilowatt-hour.)

CHAPTER FOUR

THE ADVANTAGES AND DISADVANTAGES OF WIND POWER

As is the case with any energy source, there are both advantages and disadvantages to wind power. It offers a clean, readily available alternative to fossil fuels but does not work in all locations, requires a lot of space, and has met with resistance from local residents.

WHY WE NEED WIND POWER

At the beginning of the twenty-first century, about 80 percent of the world's energy supply was derived from fossil fuels such as coal, petroleum, and natural gas. Fossil fuels are finite resources. Most estimates suggest that the proven reserves of oil are large enough to meet global demand at least until the middle of the twenty-first century. Fossil fuel

This coal-burning power plant is in Poland. Most countries around the world are now working to reduce their dependence on fossil fuels.

combustion has a number of negative environmental consequences. Fossil-fueled power plants emit air pollutants such as sulfur dioxide, particulate matter, nitrogen oxides, and toxic chemicals (heavy metals such as mercury, chromium, and arsenic). Mobile sources, such as fossil-fueled vehicles, emit nitrogen oxides, carbon monoxide, and particulate matter. Exposure to these pollutants can cause heart disease, asthma, and other human health problems. In addition, emissions from fossil fuel combustion are responsible for acid rain, which has led to the acidification of many lakes and consequent damage to aquatic life, leaf damage in many forests, and the production of smog in or near many urban areas. Furthermore, the burning of fossil fuels releases carbon dioxide (CO_2), one of the main greenhouse gases that cause global warming.

GLOBAL WARMING

The rise of air temperatures near Earth's surface over the past one to two centuries is known as global warming. Earth has experienced periods of gradual warming and cooling throughout its existence as a result of natural causes. Such natural causes have included volcanic eruptions and variations in the sun's output. However, scientists have attributed the recent increase in global temperatures to human causes—primarily the release of certain gases into the atmosphere as a result of industrial activity. These gases are known as greenhouse gases. They absorb and trap heat emitted from Earth's surface through a process known as the greenhouse effect.

While burning fossil fuels uses them up, harnessing wind's kinetic energy doesn't keep it from blowing. Wind is one of the basic elements of Earth's climate, and the natural forces that lead to its formation—the sun shining, the behavior of warm and cold gases—aren't going to cease to exist any time soon. Nor does wind power produce pollution. Since nothing is being burned, there are no inconvenient products of that chemical reaction to deal with.

THE DOWNSIDES OF WIND POWER

Wind farms are most cost-effective in areas with consistent strong winds. However, these areas are not necessarily near large population centers. Thus, power lines and other components of electrical distribution systems must have the capacity to transmit this electricity to consumers. In addition, since wind is an intermittent and inconsistent power source, storing power may be necessary.

A major concern of wind turbine siting relates to negative environmental impacts associated with noise, visual disturbance, and impacts on wildlife. Two kinds of noise associated with turbines are mechanical noise, which is produced by its equipment such as its gearbox, and aerodynamic noise, which is produced from the movement of air over the blades.

Wind farms, such as this one in Spain, tend to be in somewhat remote locations. This means they need a reliable network of power lines connecting them to the power grid.

Mechanical noise may be dampened by altering mechanical components of turbines. Aerodynamic noise, often described as a "swishing" sound, is a factor of types of blades and speed of rotation. Wind turbine noise in decibels, however, has been found to be no louder than that experienced by traveling in a moving car and is often comparable to nighttime rural background noise. Other concerns involve flicker zones, where light may be reflected off the spinning blades, and pockets of electromagnetic interference that affect television and radio signals within close proximity to turbines.

Some of the largest concerns with wind turbine placement are found in public perceptions of their visual impact and concerns about the return on investment in wind developments. For example, much controversy surrounded the 130-turbine, 468-megawatt Cape Wind project off the coast of Massachusetts, which was approved for development in April 2009 after an eight-year federal review. Located in Nantucket Sound, the project drew opposition centered on potential negative aesthetic effects the wind farm might have on scenic vistas within range of tourist destinations and second homes along Cape Cod. In April 2010, following nine years of regulatory review of the Cape Wind Project, the U.S. Department of

To help gauge the visual impact of offshore wind turbines, this photograph was prepared with images of a typical wind turbine modified to show its appearance at various distances from the shoreline.

the Interior announced its approval for the wind farm to be built. However, continued local opposition and other issues ultimately killed the venture.

Wind turbines have also been associated with killing birds at notable wind farm locations such as Altamont Pass, California. However, it is estimated that one or two birds per turbine per year are killed by wind turbines, with the majority of turbines having no impact at all. The National Audubon Society, a large environmental group based in the United States and focused on the conservation of birds and other wildlife, is strongly in favor of wind power, provided that wind farms are appropriately sited to minimize the impacts on migrating bird populations and important wildlife habitat.

This flock of Eurasian cranes is flying past a wind turbine in Diepholz, Germany. Engineers are working on ways to make wind turbines more friendly to birds and other wildlife.

However, a much higher number of bats were reported to have been killed by wind turbines. While the exact cause of those fatalities is unknown, the migration and mating behavior of migratory tree bats is widely discussed and is currently being researched by biologists.

GLOSSARY

ACIDIFICATION Making something more acidic.

AERODYNAMIC Relating to the motion of fluids (as air) that are gases and with the forces acting on bodies exposed to them.

AESTHETIC Of or relating to beauty or what is beautiful.

ANNULAR Of, relating to, or forming a ring.

ASCENDING Climbing or rising up.

CARBON DIOXIDE A heavy colorless gas (CO_2) that is formed especially by the burning and breaking down of organic substances (as in animal respiration), is absorbed from the air by plants in photosynthesis, and has many industrial uses.

COMBUSTION A chemical reaction that produces heat and light, such as fire.

CONDENSE To change from a gas to a liquid.

DIAMETRICALLY In a way that is the complete opposite.

ELECTROMAGNETIC INTERFERENCE When an external source of electromagnetic radiation (such as microwaves or radio waves) disrupts the flow of electricity through a circuit.

EMISSIONS Things that are released, particularly pollutants.

EVAPORATION The changing of a liquid into a gas.

GENERATOR A machine by which mechanical energy is changed into electrical energy.

GLOBAL WARMING A warming of Earth's atmosphere and oceans that is widely thought to result from an increase in the greenhouse effect caused by air pollution.

INEXHAUSTIBLE Unable to be used up.

INTERMITTENT Starting, stopping, and starting again.

INTERNAL-COMBUSTION ENGINE An engine which is powered by the burning of fuel, such as gasoline.

IRRIGATION Setting up a system to bring water to plants.

KINETIC ENERGY Energy associated with motion.

PARTICULATE A substance made up of very small separate particles.

ROTOR A part that rotates in a stationary part (as in an electrical machine).

SHAFT A commonly cylindrical bar used to support rotating pieces or to transmit power or motion by rotation.

SMOG A thick haze caused by the action of sunlight on air polluted especially by smoke and automobile exhaust fumes.

TURBINE An engine whose central driving shaft is fitted with a series of blades spun around by the pressure of a fluid (as water, steam, gas, or wind).

FOR MORE INFORMATION

Centore, Michael. *Renewable Energy* (North American Natural Resources). Broomall, PA: Mason Crest, 2016.

Dickmann, Nancy. *Harnessing Wind Energy* (Future of Power). New York, NY: PowerKids Press, 2017.

Doeden, Matt. *Finding Out about Wind Energy* (Searchlight Books: What Are Energy Sources?). Minneapolis, MN: Lerner Publications Company, 2015.

Hirsch, Rebecca E. *Birds vs. Blades?: Offshore Wind Power and the Race to Protect Seabirds*. Minneapolis, MN: Millbrook Press, 2017.

Jacobson, Ryan. *How Wind Turbines Work* (How Things Work). Mankato, MN: Child's World, 2012.

Kopp, Megan. *Energy from Wind: Wind Farming* (Next Generation Energy). New York, NY: Crabtree Publishing, 2016.

Macaulay, David. *The Way Things Work Now*. Boston, MA: Houghton Mifflin, 2016.

Morlock, Theresa. *Wind Farms: Harnessing the Power of Wind* (Powered Up! A STEM Approach to Energy Sources). New York, NY: PowerKids Press, 2018.

Nardo, Don. *Careers in Renewable Energy*. San Diego, CA: ReferencePoint Press, 2018.

Otfinoski, Steven. *Wind, Solar, and Geothermal Power: From Concept to Consumer.* New York, NY: Children's Press, 2016.

Owen, Ruth. *Energy from the Wind: Generating Power with Wind Turbines* (Power: Yesterday, Today, Tomorrow). New York, NY: PowerKids Press, 2013.

Small, Cathleen. *Wind, Waves, and the Sun: The Rise of Alternative Energy* (The History of Conservation: Preserving Our Planet). New York, NY: Cavendish Square Publishing, 2018.

Spalding, Maddie. *The Science of Wind Energy*. San Diego, CA: ReferencePoint Press, 2018.

Woelfle, Gretchen. *The Wind at Work: An Activity Guide to Windmills*. Chicago, IL: Chicago Review Press, 2013.

Woll, Kris. *Wind Energy* (Alternative Energy). Minneapolis, MN: Core Library, 2017.

WEBSITES

International Energy Agency

https://www.iea.org/topics/renewables/subtopics/wind

Facebook: @internationalenergyagency; Twitter: @IEA

National Renewable Energy Laboratory
https://www.nrel.gov/workingwithus/re-wind.html
Facebook: @nationalrenewableenergylab
Twitter: @nrel

U.S. Department of Energy's Office of Energy Efficiency and Renewable Energy
https://energy.gov/eere/wind/wind-energy-technologies-office
Facebook: @energygov; Instagram, Twitter: @energy

Wind Energy Foundation
http://windenergyfoundation.org
Facebook: @WindEnergyFoundation;
Twitter: @windenergyfdn

INDEX